Manual of Water Supply Practices

M72

# Knife Gate Valves

## First Edition

**AWWA Reference Standard C520-19**

AF904081

American Water Works Association

Manual of Water Supply Practices—M72, First Edition

# Knife Gate Valves



If you find errors in this manual, please email books@awwa.org. Possible errata will be posted at www.awwa.org/M72.

Managing Editor/Project Manager: Melissa Valentine
Technical Editor: Dianne B. Rose
Cover Design: Mike Labruyere
Manager, Publishing Operations: Gillian Wink
Production: Innodata
Senior Specialist - Manuals: Willadee Hitchcock
Manuals Coordinator 2: Janet Greifinger

---

**Library of Congress Cataloging-in-Publication Data**

Names: MacConnell, Mark, author.
Title: M72 - knife gate valves / Mark MacConnell.
Other titles: Knife gate valves | AWWA manual ; M72.
Description: First edition. | Denver, CO : American Water Works Association, [2020] | Series: Manual of water supply practices ; M72 | Includes bibliographical references and index. | Summary: "This manual presents the general practice for selection and installation of manual and automated knife gate valves for use in water and waste-water applications. This document is intended to provide information and guidance on typical knife gate valves and their intended application"-- Provided by publisher.
Identifiers: LCCN 2020045228 (print) | LCCN 2020045229 (ebook) | ISBN 9781647170271 (paperback) | ISBN 9781613005675 (ebook)
Subjects: LCSH: Water-pipes--Valves--Handbooks, manuals, etc.
Classification: LCC TD491 .M18 2020 (print) | LCC TD491 (ebook) | DDC 628.1/5--dc23
LC record available at https://lccn.loc.gov/2020045228
LC ebook record available at https://lccn.loc.gov/20200452299

Printed in the United States of America

ISBN 978-1-64717-027-1 eISBN 978-1-61300-567-5
DOI https://doi.org/10.12999/AWWA.M72ed1



American Water Works Association

American Water Works Association
6666 West Quincy Avenue
Denver, CO 80235-3098
**awwa.org**

# Contents

# Figures

This page intentionally blank.

# Tables

This page intentionally blank.

# Preface

This manual presents the general practice for selection and installation of manual and automated knife gate valves for use in water and wastewater applications. This is a manual of recommended practice, not an AWWA standard. This document is intended to provide information and guidance on typical knife gate valves and their intended application. Questions about specific installations or applicability to specific applications not addressed in this manual should be directed to the manufacturer or supplier.

Information contained in this manual is useful for operators, technicians, and engineers to gain a basic understanding of the use and application of manual and automated knife gate valves. There are many variations of knife gate valves that may exceed the scope of this manual and require special consultations with the manufacturer to determine the associated application requirements. The valves' operational and design information is generic, and actual data from the intended valve manufacturer should be obtained before making the final selection of any type of knife gate valve and its options. This manual supplements information on the knife gate valves listed in ANSI/AWWA Standard C520 of latest edition: Knife Gate Valves, Sizes 2 in. (50 mm) Through 96 in. (2,400 mm).

AWWA standards are available for purchase by calling (800) 926-7337 or online at https://www.awwa.org/publications.

Manufacturers graciously provided valve illustrations and other documentation. AWWA does not endorse any manufacturer's products, and the names of the manufacturers have been removed from the material provided.

Metrification Note: Valves are listed in their current US designation (in.) and the equivalent nominal metric designation (mm). To obtain the actual metric equivalent, use a conversion factor of 25.4 mm per inch.

This page intentionally blank.

# Committee Personnel

The AWWA Standards Committee on Knife Gate Valves, which developed this manual, had the following personnel at the time of approval.

Mark MacConnell, *Chair*

***General Interest Members***

Michael Altland, Mott MacDonald, Iselin, N.J.
Frank Blowers, Beaver Water District, Lowell, Ariz.
Anthony Holmes, Jones Edmunds & Associates, Inc., Winter Haven, Fla.
Christine Kirby, Lockwood, Andrews & Newnam, Inc., Houston, Tex.

***Producer Members***

David Gambetta, Lined Valve / Henry Pratt / Mueller, Woodland, Wash.
Paul Gifford, Mueller Co., Chattanooga, Tenn.
Jerome Grant, DeZURIK, Sartell, Minn.
Bruce James, Iconix Waterworks, Milton, Ont.
Al Libke, DeZURIK, Sartell, Minn.
Michael Long, Elite Valve, Oakville, Ont.

***User Members***

Jeffrey Archer, Denver Water, Denver, Colo.
Mark Johnson, Massachusetts Water Resources, Chelsea, Mass.
Mark MacConnell, Burnaby, B.C.
Ronald Masek, City of Houston, Houston, Tex.
Jim Young, City of Richmond, B.C.

This page intentionally blank.

Chapter 1

# Introduction

AWWA knife gate valves are a versatile design manufactured in wafer, lug, and flanged body configurations from various materials. Designs are available with nonadjustable packing in bonneted-style valves and adjustable packing glands in nonbonneted-style valves. AWWA designs provide superior performance and reliability over conventional valve designs because of requirements for low- and high-pressure valves used in various applications associated with the waterworks industry. Knife gate valves are used in liquid and suspended solids handling. They are also used for flow isolation as well as throttling/control duties under certain operating conditions in modified configurations available from some manufacturers.

Most knife gate valve applications involve relatively low pressures, typically 150 psig (10 bar) or less; however, special designs are used up to 300 psig (20 bar) and higher (e.g., ASME Class 300). Because of some standard constructional features, knife gate valves in metal-seated configurations and selected packing material can handle major temperature fluctuations that may seriously damage or hamper the function of other valve designs or resilient-seated valves.

## HISTORY

Simple gate valves date back to ancient times in Greek and Roman water and irrigation systems. These gates were mostly operated by manual levers or screws.

The first modern knife gate valve, simply known as the gate valve, was produced in Germany in 1881. Knife gate valve mass production started around 1928, primarily for the pulp and paper industry. The name "knife" gate valve has been used in conjunction with the thinner, round, beveled leading edge of the gate to distinguish this valve from the common wedge gate valve. The first generation of knife gate valves came with a bottom-draining pocket. Soon after, standard valves from most manufacturers were made without a bottom pocket and only in rising stem configurations. The name "pulp valve" was commonly associated with today's knife gate valve, which at the time was available in diameters of 3 in. (75 mm) up to 24 in. (600 mm).

During the initial manufacturing years of the early 1900s, the valve was available only with cast iron bodies. Leakage was common in these earlier models, which only employed metal-seated configurations. (Newer models now offer unidirectional and bidirectional resilient seating at low and high pressures with zero leakage capabilities.)

A degree of control functionality was then incorporated into the valve. Many knife gate valves are operated with lever, handwheel, chainwheel, and bevel gear manual actuators (see Figure 1-1). Powered actuators, either cylinder or electric, are also used regularly in many industrial applications for knife gate valves, including the paper, oil, gas, and power industries.

## PURPOSE

This manual will explain fundamental concepts and operation of knife gate valves, providing users with a technical background to understand, specify, and install knife gate valves in applications for which they are intended. The information in this manual should be used in conjunction with data from valve manufacturers to select valves and their associated actuation for a given application.

For isolation and control applications, actuator sizing parameters such as thrust will be explained in detail so that the sizing methodologies recommended by valve manufacturers can be applied. It is also important to understand how certain valve

*Source:* Elite Valve

**Figure 1-1 Typical bonnetless knife gate valve**

application parameters such as flow rate, differential pressure, and direction of flow affect actuator selection and operation. Changes to the flow parameters after the selection and installation of the valve and actuator may affect the actuator's ability to properly operate the valve.

General installation guidelines will be presented to provide users with a basic understanding of startup procedures and enable them to troubleshoot certain valve installation issues, ensuring proper and effective operation of the valve.

## BODY STYLE

Many different body styles are available. The type of service for which the valve is suited will directly influence the valve's body style as well as the connection methodology to piping systems. In all styles presented herein, there are two or more blind tapped holes in the body chest area to maintain metal seal integrity to the internal packing chamber. Shorter flange bolts will be required in these locations.

### Wafer

This is installed by compressing the slender valve body between the gaskets of two adjoining flanges. This body style does not have any threaded holes to hold the valve in place between the flanges but can feature bored-out locating holes for adjoining flange through bolts around the perimeter. Better pipeline sealing characteristics are obtained when the valve body contains raised faces in conjunction with mating raised face flanges. It is not recommended that valves with raised faces be used to mate with flat-faced, ring-type flanges as the flat-faced flange may not be able to withstand the forces placed on its outermost edge across the gap, and the sealing characteristics may be detrimentally affected. This valve cannot be used for end-of-line or dead-end service.

### Lug

This valve body style comes with a fully threaded bolting pattern through the body lugs that are located radially around the outside perimeter of the valve (see Figure 1-2). The lug

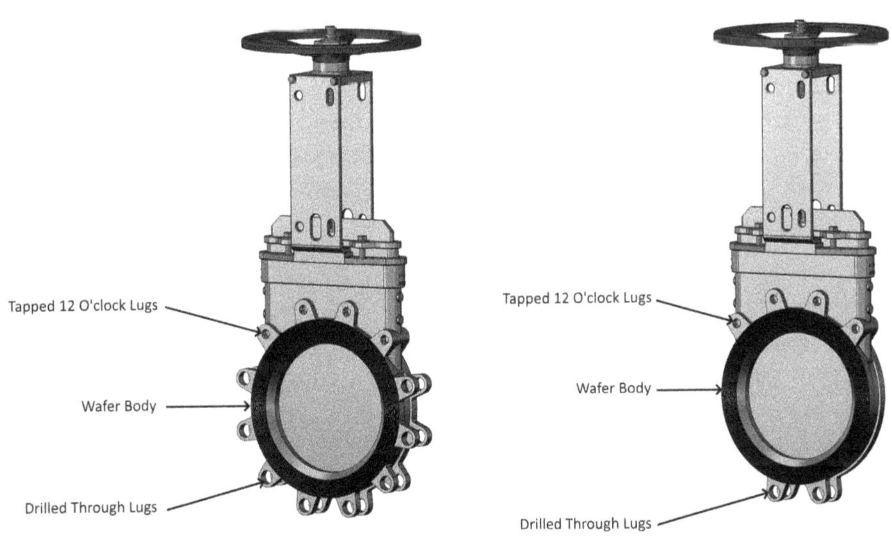


*Source:* Elite Valve

**Figure 1-2 Typical bonnetless knife gate valve with drilled-through bolt pattern (left) and tab-less wafer body (right)**

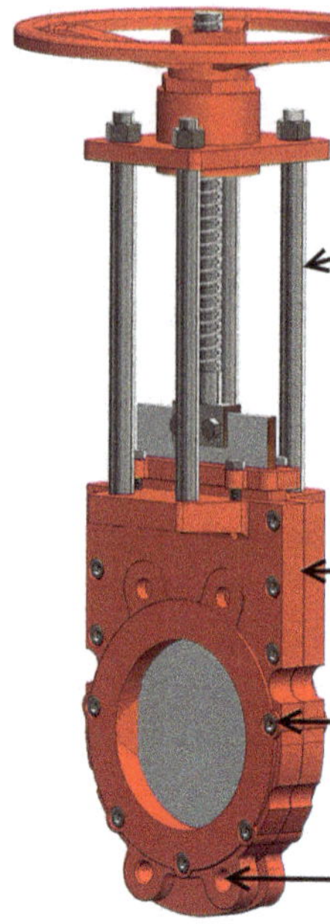


*Source:* Elite Valve

**Figure 1-3 Semi-lug bonnetless knife gate valve with two-piece body and stanchions**

and bolt pattern can vary to match adjoining companion flanges based on the required flange standard and design pressure classifications of the system component or facility. This valve style is ideal for end-of-line service or where it may be necessary to disconnect one side of the valve from the adjacent piping without having to depressurize the opposite side of the valve.

## Semi-Lug

This body style is a cross between the wafer and lug styles (see Figure 1-3). It features threaded lugs through the valve body and around the perimeter of the valve. Typically, only the 12 o'clock and 6 o'clock lug positions are threaded. As a result of the lug locations, this valve can be removed without complete disassembly of the mating piping. It is not recommended for end-of-line service because of the reduced number of fastening bolts, which lends this style to rely on the adjoining companion flanges and rigid piping system to resist the pressure and thrust forces when closed.

## Flanged

Flanged valves have a longer face-to-face dimension than wafer- or lug-style valves and may not be interchangeable without having to modify either the upstream or downstream adjacent piping (see Figure 1-4). Depending on the piping configuration, this may or may not be possible. The bolt holes are full bore through the flange portion of the valve body matching the requirements of the applicable flange standard and are not threaded. This valve style is ideal for end-of-line service or where it may be necessary to disconnect one side of the valve from the adjacent piping without having to depressurize the opposite side of the valve.

*Source*: Elite Valve

Figure 1-4 Typical bonneted double-flanged knife gate valve

## GENERAL APPLICATION

In its correct form and configuration, a manual or automated knife gate valve is ideal for use in many applications handling water, wastewater, viscous media, dry powder, pellets of various materials, slurries, sludges, mine tailings, and food, as well as for use in the cement, brewing, and pulp/paper industry.

Most knife gate valve manufacturers offer the valves in rising and non-rising stem configurations that are actuated with a lever, handwheel, chainwheel, pneumatic cylinder, hydraulic cylinder, or electric motor.

For water and wastewater uses, a knife gate valve can be configured for exposed, buried, submerged, and vaulted applications. In buried and submerged applications, the addition of the pressure bonnet—which encapsulates the entire gate during valve cycling—will keep the gate within the pressure boundary and eliminate the packing seal around the gate. In these installations, it is recommended to install a sleeve surrounding the stem of the knife gate valve from the top of the bonnet up to the surface to protect the stem and avoid interference with the stem during cycling operations. There are various types of actuation that can offer different options for indicating the valve's operational position.

The knife gate valve can be designed for applications where the flow is applied to the gate in either a unidirectional or bidirectional service condition. Valve designs for various applications include metal- or resilient-seated configurations. Resilient-seat designs can offer zero leakage performance at the manufacturer's specified pressures. However, in metal-seated valve configurations, an acceptable manufacturer-specified leakage rate is typically permitted. In certain applications, knife gate valve designs may require minimum pressures to ensure zero leakage, so requirements must be reviewed before selection of a design for the intended valve application. This information is found in either the manufacturer's product literature or in their installation, operation, and maintenance manual. In all cases, both documents from the manufacturer should be reviewed before the selection of a valve for any application to ensure the anticipated performance meets the criteria of the intended application.

**Table 1-1 Typical service application requirements**

| | Wafer | Full Lug | Semi-Lug | Flanged |
|---|---|---|---|---|
| **End-of-line Removal** | XX | √√ | X | √√ |
| **Removal from Fixed Pipework** | √√ | √ | √ | X |
| **Interchangeable with Other Body Styles** | √√ | √√ | √√ | XX |
| **Pressure Isolation** | √ | √√ | √ | √√ |

*Source:* Elite Valve
√√ - Recommended for use
√ - Acceptable for use
X - Not recommended for use
XX - Do not use

**Table 1-2 Typical applications and characteristics for various valves**

| | Knife Gate Valves | Wedge Gate Valves | Butterfly Valves | Ball Valves | Plug Valves |
|---|---|---|---|---|---|
| **Stringent Solids Handling** | √ | X | X | X | √ |
| **Throttling** | X | X | √ | √ | √ |
| **Full Port Flow** | √ | √ | X | √ | √ |
| **Seal Options** | √ | X | √ | X | √ |
| **Replaceable Seals** | √ | X | √ | √ | X |

*Source:* Elite Valve
√ - Recommended
X - Not recommended

Table 1-1 indicates the typical in-service application requirements for the different knife gate body styles.

Depending on the design and configuration of the internal porting, these valves can be used in applications where piping systems require pigging or swabbing. Dimensions must be validated by the manufacturer for use in these applications.

In most body designs, the valve face is either flat or raised depending on the valve's pressure rating. Flat-faced valve bodies are typically used in low- to medium-pressure applications, while raised-face valve bodies are generally used in high-pressure applications. The gasket type and material to be used should be verified with the corresponding valve or pipe flange rating and dimension for each application, as the gaskets will be exposed to the pipeline media when using knife gate valves. Certain valve designs may incorporate O-ring seals or special inserts directly into the flange face, in which case flange gaskets may not be required.

Knife gate valves with adjustable packing glands or bonnetless configurations are not designed or recommended for buried or submerged service applications.

Table 1-2 shows various types of valves and associates typical characteristics with their recommended applications, operation, and maintenance processes.

Typically, valve costs are directly related to weights of the valve. Given the design of significantly heavier wedge gate, butterfly, ball, and plug valves, knife gate valves are typically less expensive as they have a narrower design profile than the other valve types. This difference in cost becomes more significant as the valves increase in size.

Knife gate valves are typically designed to be full bore or reduced bore. Reduced-bore designs are sometimes referred to as standard bore and are typically 90 percent of the inside diameter of Schedule 40 pipe. This large, unobstructed flow area allows an increased ability for laminar flow with minimal head loss when compared to other valve types such as butterfly, ball, and plug. The manufacturer or representative can identify which valve models are either standard or full bore.

This page intentionally blank.

Chapter 2

# Principles of Operation

This chapter applies to the types of knife gate valves used in the water and wastewater industry described in ANSI/AWWA C520, Knife Gate Valves, Sizes 2 in. (50 mm) Through 96 in. (2,400 mm), of latest edition. The operation of these valves in their intended application will be explained, as well as their construction and function.

## DEFINITIONS

A knife gate valve has a thin, flat gate with either a relatively sharp, beveled semicircular leading edge for metal-seated valves or a rounded, semicircular leading edge for rubber-seated valves that moves or operates within the valve body in relation to a stationary seat when actuated with a manual or power-operated device.

Figure 2-1 and Figure 2-2 illustrate the components typical of knife gate valves.

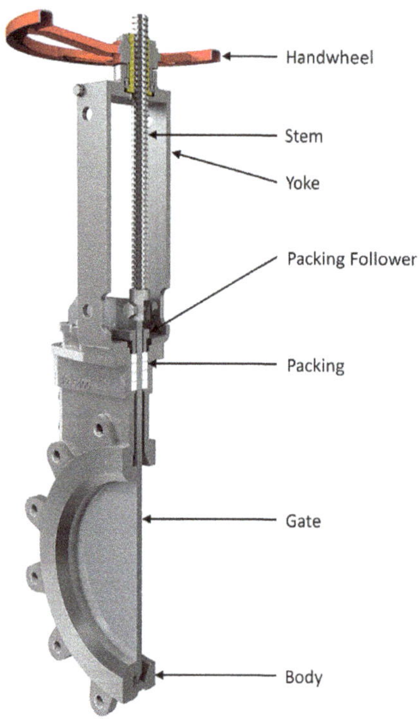


*Source:* Elite Valve

Figure 2-1 Cross section of a typical bonnetless knife gate valve with two-piece yoke and handwheel

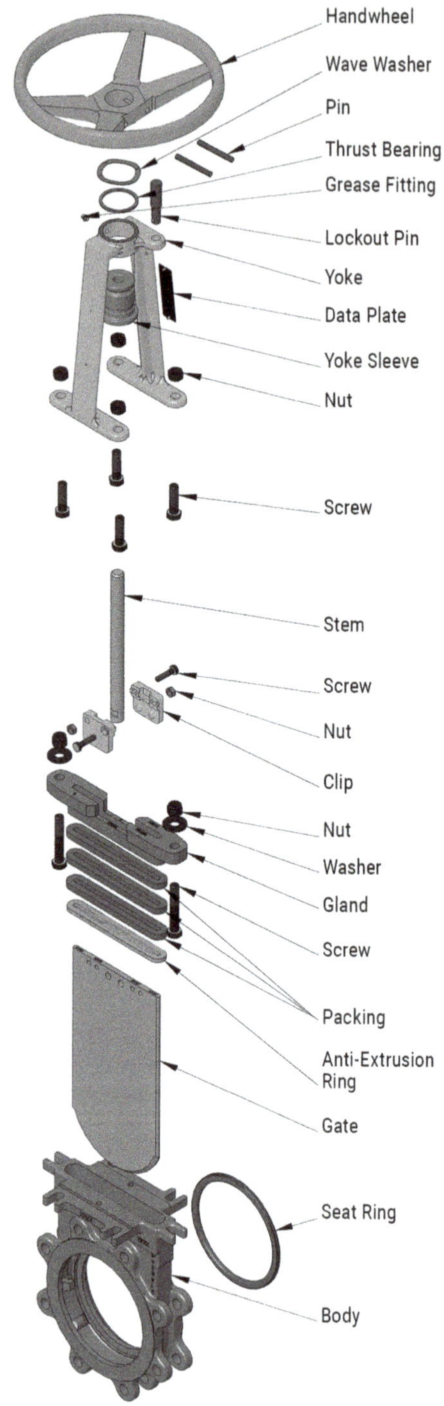


*Source:* DeZURIK Inc.

Figure 2-2 Exploded view of a typical bonnetless knife gate valve

## FUNCTION

In most cases, the gate moves to close the valve with a clockwise rotation of the handwheel. In some cases, when requested, a clockwise rotation of the stem may be provided to open the valve. Knife gate valves operate in a linear motion within the body, which allows fluids to flow through the opening at varying rates.

Knife gate valves can be used for any liquid and many slurries; however, to maximize the valve's capacity and its resiliency, the appropriate size must be selected for the given application. The common installations for these valves include when high flow rates are anticipated; unobstructed full port flow is required; where thick or viscous sludges may be in the system such as in wastewater treatment systems; and/or in tight piping configurations where space is limited.

Knife gate valves can also be furnished with special end connections for use as tapping valves. Tapping knife gate valves require an unobstructed waterway and a special flange face to provide proper alignment and position of the tapping sleeve and machine.

This page intentionally blank.

Chapter 3

# Types of Construction

This chapter presents the typical designs of knife gate valve construction, which includes body styles and types of construction, sealing mechanisms, principle valve components, seating configurations, and methods of seating. With an understanding of the types and advantages of different seating configurations, the user will be able to apply the best configuration for a given application. This manual does not present all available types of knife gate valves, but it generally covers those contained in the AWWA C520 standard for knife gate valves.

## GENERAL CONSTRUCTION

The gate or knife is a device with a flat plate usually ground and polished on both faces that is internally retained in the valve body. The body configuration is typically narrow in wafer, lug, or double-flanged designs, requiring minimal face-to-face laying length in piping systems. The valve may have a bonnet to assist in containing the fluid and vapor emissions within the pressure boundary of the valve in high-pressure or direct buried applications. It may also be bonnetless, which is typically found in nonburied applications where lower pressure flow conditions exist. However, some bonnetless designs have options that can accommodate higher-pressure applications.

The user should be aware that bonneted design valves are costlier, heavy, and have significantly higher centers of gravity than that of bonnetless designs.

## BODY CONSTRUCTION

The type of body construction determines how the valve is assembled and can affect the types of options and accessories available.

### Cast Single Bodies

Available in various grades of cast iron, cast steel alloys, and stainless steels. These valves can accommodate either unidirectional or bidirectional flow conditions.

### Cast Bolted Bodies

Sometimes referred to as a "split body" design, where the valve body is cast as two separate pieces and then fastened together with screws during assembly. These valves can also accommodate either unidirectional or bidirectional flow conditions.

### Fabricated

Fabricated valves are often of low-volume production quantities in special sizes, rectangular shapes, and other nonstandard valve designs (as well as using special alloys) that typically make use of forged or rolled plate with certified welding techniques and nondestructive weld examinations. These valves can also accommodate either unidirectional or bidirectional flow conditions.

## GATE PACKING AND STEM SEALS

Located around the upper portion of the main body where the gate exits for bonnetless valves, several layers of packing in the packing gland surround the gate and maintain the process pressure by containing the media within the internal pressure boundary of the valve. For bonneted knife gate valves, a stem seal is located in the bonnet around the stem to maintain process pressure and prevent media from escaping the internal pressure boundary of the valve.

### Bonnetless Gland Packing

Typical packing consists of layers of braided cord that are compressed tightly against the shape of the gate and packing gland by a packing follower gland bolted to the main valve body, which allows the gate to move within the valve body while maintaining a positive seal. The braided cord is available in a variety of materials; polytetrafluoroethylene (PTFE) is the most common. Selection of other packing material for compatibility with pressure, temperature, and/or special media can be accommodated.

### Bonneted Adjustable Packing

Bonneted knife gate valves (see Figure 3-1) do not rely on a seal around the gate but instead create a seal where the round stem protrudes through the bonnet. The seal can either be adjustable, using a design similar to the bonnetless packing gland, or nonadjustable, such as with an O-ring style seal or static packing. Although some manufacturers provide a back-seating ring for replacing packing while in service, this is not recommended for valves subjected to pressurized flow.

### Bonnetless and Bonneted Live Loaded Packing

Live loaded packing (see Figure 3-2) relies on spring-loaded Bellville-type washers or a resilient material core within packing rope to "energize" the packing when compressed. This type of packing provides better sealing characteristics when higher valve-cycle life is required.

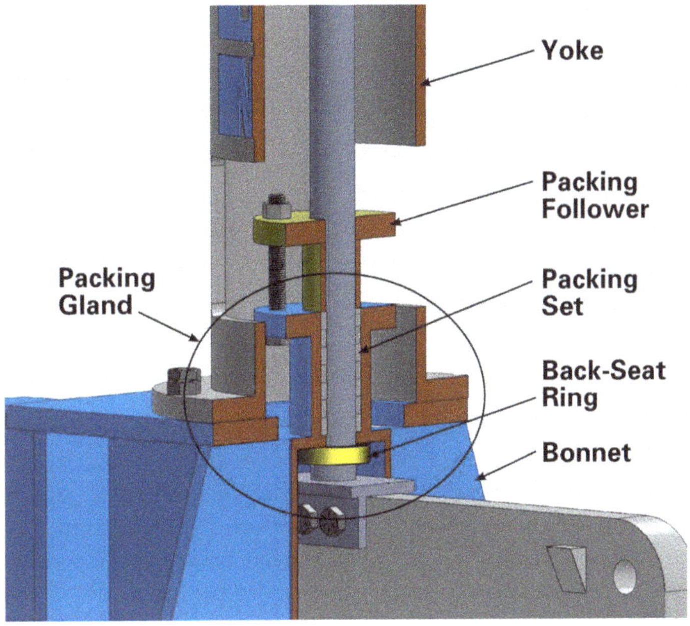


*Source*: DeZURIK Inc.

Figure 3-1 Bonneted knife gate valve with adjustable packing

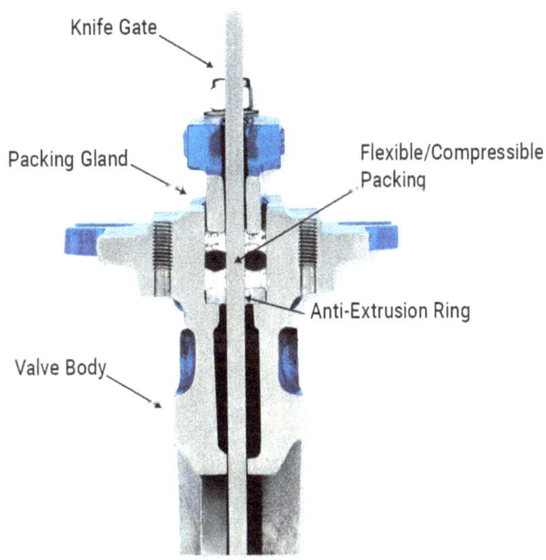


Source: DeZURIK Inc.

Figure 3-2 Bonnetless knife gate valve with live loaded packing

## Bonnetless Injectable Packing

Bonnetless injectable packing makes use of a rubber seal stretching transversely across each side of the gate. The rubber seal is forced tight against the gate by a pliable PTFE putty injected between the valve body and the rubber seal.

### Bonneted O-Ring Seal

Bonneted valves can make use of a single or double O-ring seal at the stem in both rising and non-rising stem knife gate valves. The O-ring seal is exposed to the pipeline fluid; therefore, this type of seal selection needs to consider compatibility with temperature, entrained solids, chemical, and chloramine resistance. The shaft seal will be a translational seal in the case of a rising stem valve or a rotational seal in the case of a non-rising stem valve. The shaft seal is typically the same material as seat material.

### Bonnetless High-Performance Packing

High-performance packing designs for bonnetless valves can provide an extended service life in critical service applications and specific environments. These designs typically have a packing chamber machined into the body of the valve. Antiextrusion rings having tight tolerances are inserted between the lower packing ring and the bottom of the packing chamber. This type of packing design requires the gate edges to be rounded and polished to ensure the optimum effective seal for maximum packing life with minimal or no adjustment required. This packing can also include a rubber O-ring material that becomes compressed and energized when the packing follower is tightened to compress the packing chamber. As the packing wears over time, the rubber O-ring material relaxes and maintains packing rope compression against the gate.

Packing material selection is critical for maintaining the pressure boundary of most knife gate valves. Packing materials are available for a wide range of chemical compatibility, temperature resistance, high pressure, extended service life, etc. The user should take precautions to ensure that the packing material used is compatible with the application.

## YOKES

The yoke assembly holds the actuator in place above the valve body and provides a centered alignment for extended valve stems. The yoke design is manufacturer-dependent and designed to accommodate any installed actuator.

### Pillars

Sometimes referred to as stanchions, four posts mounted directly to the body of the valve and the adaptor plate support the actuator. Pillar-style yokes are best used for linear-type (i.e., nontorqueing) actuation, such as in-line pneumatic and hydraulic cylinders, as they have much less available cross-sectional area to resist the radial forces from the actuator's torque output.

### Single Piece

Typically cast, but sometimes formed. A single, solid piece of material with a centered cavity for the valve stem supports the actuator above the valve body. Single piece-style yokes are suited for torqueing-type actuators, as they can minimize the amount of radial deflection induced by the actuator's output torque.

### Twin Piece

A twin yoke consists of two separate plate supports on opposite sides from each other centered above the valve body. Twin piece-style yokes are best used for linear-type (i.e., nontorqueing) actuation such as pneumatic or hydraulic cylinders, as they have less available cross-sectional area to resist the radial forces from the actuator's output torque.

### Bonnet

On some designs, the bonnet can replace the yoke assembly. When this style is supplied, the valve's actuator can be mounted directly to the top of the bonnet without the need for an additional yoke mechanism. Yoke mechanisms replaced by bonnets are well suited for torqueing-type actuators, as they can eliminate the radial deflection associated with the actuator's output torque.

## STEM

Stems are typically made from stainless steel materials, although high-alloy or nickel copper materials may be used for specific applications. For manual actuation, they are usually machined with an acme-style drive thread, either single or double start, and left-hand pitch. The left-hand pitch allows the valve gate to close with a clockwise turning of the actuator/handwheel and to open with a counterclockwise turning motion.

### Rising Stems

Bonnetless knife gate designs typically employ a rising stem to operate the valve gate. The stem moves in a linear direction and is not able to rotate as it is directly coupled to the gate clip/clevis.

### Non-rising Stems

Nonrising stem knife gate valves can be found in both bonnetless and bonneted designs. In this type of design, the stem is able to rotate freely, but it remains stationary and is prevented from moving linearly with the valve gate. To achieve this type of noninterfering mechanical movement, the valve stem must be offset to one side of the gate and make use of a 90-degree clevis gate attachment.

## GATES

Valve gates are typically constructed from steel plate, which is readily available in a wide variety of materials. Stainless steel materials are most commonly used for water service as indicated in AWWA C520. Gate designs and surface finishing vary depending on the intended application and in-service pressure requirements. The importance of valve gate design becomes more significant as valve sizes and pressures increase given the thinner blade style inherent with knife gate valves. Gate surface finishing becomes more critical as the requirements for valve performance increase. The edge preparation of valve gates is important to ensure the valve packing and seats remain undamaged from regular operation or long periods of inactivity.

## VALVE SEAT CONFIGURATIONS

The valve seat provides the seal for process isolation between the upstream and downstream sides of the valve. Knife gate valves are designed for either unidirectional or bidirectional sealing capabilities. In either type of seat design, the valve may be capable of either positive drip-tight, zero-leakage seal, or a specified amount of leakage may be permissible.

### Unidirectional Seat Design

Unidirectional seat designs are typically either in the form of a metal or resilient seat where the body of the valve has a single seating side (see Figure 3-3).

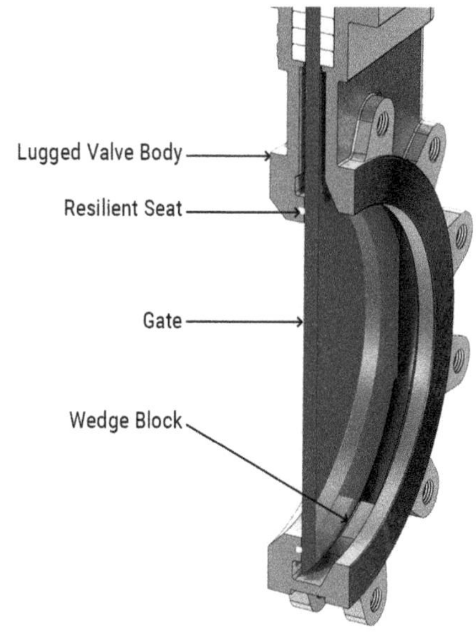


*Source*: Elite Valve

**Figure 3-3 Resilient-seated knife gate valve**

**Metal-seated.** Typically used when leakage is permissible for liquids not containing any particulate, when dewatering of the pipeline is not a concern, or for clear liquid applications with flow in the preferred direction where the pressure differential is assisting the valve seal. Media dewatering can occur adjacent to the gate of closed metal-seated valves based on the media in the pipe. Exercise care when dewatering a pipe with certain types of media present in the liquid where the valve is closed against a metal seat.

**Resilient-seated.** Normally used in applications with flow in a preferred direction where zero leakage is required throughout the entire differential pressure range of the valve. Some resilient-seated valves require a minimum pressure differential to achieve zero leakage. When this is specified, the manufacturer should confirm the definition and requirements necessary to obtain zero leakage of the valve based on system media and flow parameters. It is recommended to specify the differential pressure if it is below 40 psi (276 kPa).

## Bidirectional Seat Design

Bidirectional seat designs employ a seating surface on both sides of the valve body in either a dual resilient face seal or perimeter-seated configuration (see Figure 3-4).

**Dual resilient-seated.** Normally used in applications where flow can occur in both directions when there is a requirement for zero leakage. A seal or O-ring style gasket is installed and retained inside the valve body on both seating surfaces. Typically, dual or twin resilient-seated valves *do not* require a minimum pressure differential to achieve zero leakage; however, this should be confirmed by the manufacturer as to the definition and requirements necessary to obtain zero leakage of the valve based on system media and flow parameters.

**Perimeter-seated.** Normally used in applications with flow in both directions when zero leakage is required throughout the entire differential pressure range of the valve. A single seal or gasket is installed along and against the inside seating perimeter of the valve body. The gates on some perimeter-seated valves do not have a full and complete sharply

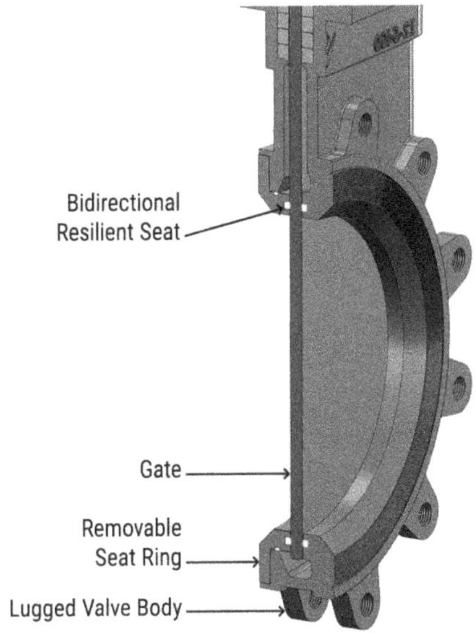


*Source:* Elite Valve

**Figure 3-4 Bidirectional resilient-seated lugged-body knife gate valve**

beveled leading edge to prevent damage to the perimeter seat. This type of seat typically requires more seating torque from the actuator to achieve zero leakage at higher differential pressures. Perimeter-seated valves *do not* require a minimum pressure differential to achieve zero leakage; however, this should be confirmed by the manufacturer as to the definition and requirements necessary to obtain zero leakage of the valve based on system media and flow parameters.

## METHODS OF SEATING

It is important to understand the seating characteristics of knife gate valves. Some types of valves are seated by applying or maintaining a downward force on the stem ("torque seated"), and other valves are seated by accurately positioning the gate or blade ("position seated").

Knife gates move in a linear direction within the valve body, whereby the closed gate is either against the flush seating surface inside the base of the body or retained within the body cavity and wedged to close in the last inch or so of travel as the gate moves into the body's fixed jams (i.e., wedges). The former seating characteristic is suitable for the position-seated design more typical of resilient-seated valves, whereas the latter seating characteristic typically pertains to metal-seated valves. A notable exception to resilient-seated designs would be the perimeter-seated valve, which may require additional linear force to effectively seat the valve. The actuator's stroke distance is typically adjusted to provide the best seating position that will prevent or minimize the amount of leakage depending on the type of valve seat. Accurate positioning is required to both seal the valve and prevent unnecessary downward force resulting in column loading on the valve stem. Metal body seats are typically machined directly into the cast body, but they may be separate, field-replaceable seat rings.

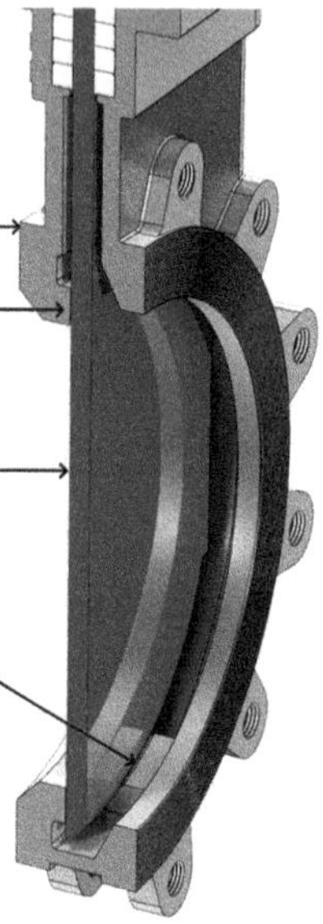


*Source*: Elite Valve

**Figure 3-5 Metal-seated lugged-body knife gate valve**

## METAL-SEATED

Metal-seated valves are normally selected when leakage is permissible and dewatering of the pipe on a closed valve can occur without any solid buildup within the valve body or in the adjacent piping (see Figure 3-5). Metal-seated designs typically have a preferred flow direction and employ a single-seat design inside valve body on one side of the valve that typically faces downstream, which is characteristic of unidirectional-flow knife gate valves. The sealing surfaces are mated and enhanced by fixed jams/wedges located in the valve body that force the gate to the closed position during the last inch or so of travel upon linear closure. Body jams/wedges are usually either cast or welded into the lower circumference of the body and machined to mate with the surface of the gate.

### Resilient-Seated

Resilient-seated valves are normally selected when zero leakage is required and/or dewatering of media inside the pipe on a closed valve is not permissible. Resilient-seated designs may be single-seated on one side of the inside valve body with a preferred unidirectional flow design or bidirectional perimeter-seated (see Figure 3-6). Some designs may have double-seated resilient elastomers that are field replaceable with retaining rings. In most cases, resilient-seated knife gate valve seats are not typically field adjustable; however, the resilient seats are field replaceable. For single-seated designs, sealing characteristics may be enhanced by fixed wedges located in the valve body that force the lower portion of the gate into the resilient seat during the last inch or so of travel upon linear closure of the valve. A perimeter resilient seat around the outside edge of the gate is typically molded in a horseshoe design and field replaceable. It can be installed in a split body where the two body halves compress the seat, holding it in place. Alternatively, it can be incorporated into a single cast body where the seat is installed through the gate channel and held in place by retaining plates, packing rope, and the packing follower.

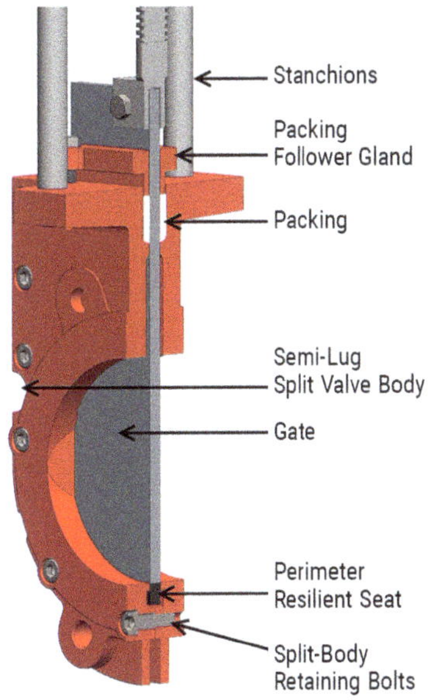


*Source*: Elite Valve

**Figure 3-6 Perimeter resilient-seated semi-lug split body knife gate valve**

This page intentionally blank.

Chapter 4

# Optional Equipment

Special designs and/or accessory equipment are available to operate the valve for on–off, modulating, and fail-safe requirements, or to provide safety lockout and feedback capability for all types of knife gate valves.

## MANUAL OVERRIDE

Pneumatic or hydraulic actuators can be supplied with a manual override to allow actuation without electrical power or stored energy—typically a lever attached to pumping mechanism. In most cases, electrically actuated valves will incorporate a manual override feature—typically a declutching handwheel (see Figure 4-1).

## LOCKOUT MECHANISM

When service or inspection is needed on a valve, piping, instruments, fittings, and/or where the pipeline must be entered, under North American safety regulations a lockout mechanism is required on valves being used as an integral component of the system lockout (see Figures 4-2 and 4-3). The knife gate valve design has the ability to provide this additional safety feature. It can prevent the operation of the valve—local or remote controlled—on manual or automated knife gate valves. Any time personnel may have the ability to reach inside an opened valve, a mechanical lockout device will provide an acceptable measure of safety against unauthorized valve operation. In particular, lockout mechanisms should be considered if the downstream piping will be dismantled from the valve, where it is automated with a quick-acting pneumatic cylinder. A lockout mechanism will also prevent an accidental opening of a valve where the potential for a release of high pressure and/or high-temperature fluid or other dangerous substance is present.

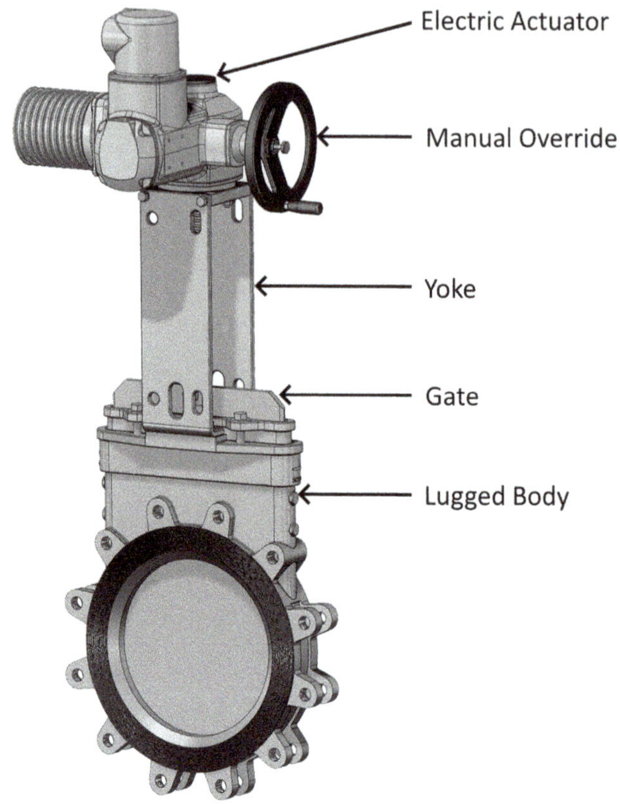


*Source*: Elite Valve

**Figure 4-1 Handwheel manual override on electrically actuated bonnetless lugged knife gate valve**

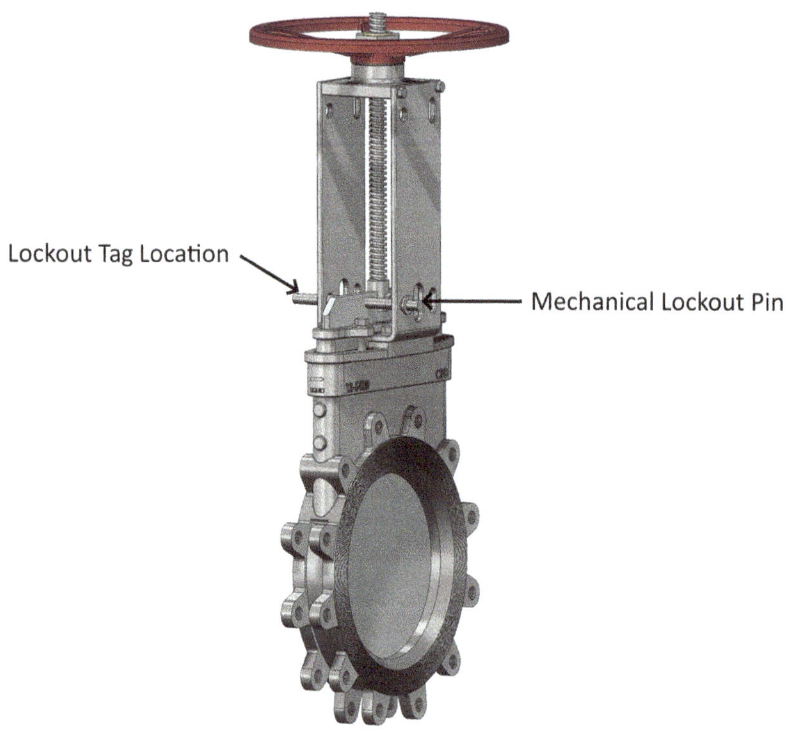


*Source*: Elite Valve

**Figure 4-2 Mechanical lockout device**

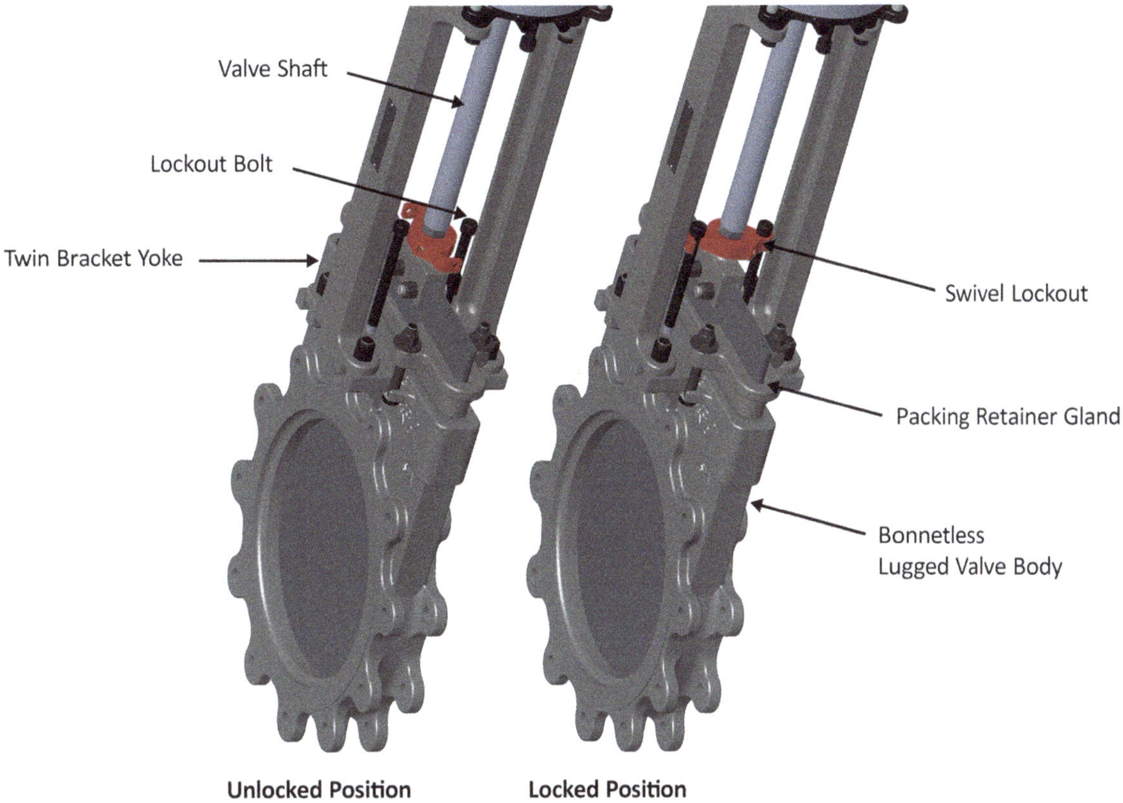


*Source*: DeZURIK Inc.

Figure 4-3 Swivel-type lockouts

To facilitate this type of safety mechanism, manufacturers have either cast holes in the handwheel to lock it in place or drilled aligned holes through the gate and yoke for the insertion of pins to prevent the gate from operating.

In all cases, these mechanical safety devices must be able to withstand the maximum output thrust force of any actuation. It is also necessary to verify the supplied actuation's maximum output thrust force with each manufacturer for compliance with and the capability of their lockout mechanism's safety design.

## EXTERNAL GATE GUARDS

State, provincial, or federal health and safety guidelines may require gate guards to be mounted on the body of all powered/automatically actuated bonnetless knife gate valves (see Figure 4-4). The purpose of the gate guards is to prevent injury if an automated knife gate valve cycles without warning.

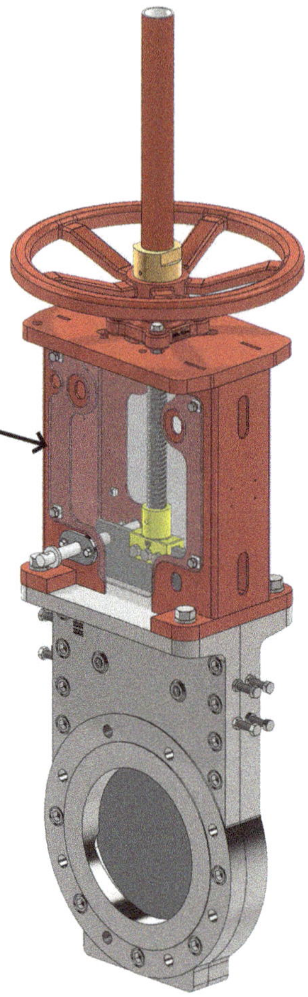


*Source*: Elite Valve

**Figure 4-4 Gate guard safety screen**

## FAIL-SAFE OPERATION

Hydraulic and pneumatic actuators can be equipped with springs or accumulators, which store energy to provide half of a valve cycle for fail-to-open, fail-to-close, or fail-on-command operation with loss of electrical power or pneumatic or hydraulic supply pressure. Solenoids used on hydraulic or pneumatic systems can be piped to operate the valve to fail open or closed on loss of power to the solenoids or on-command through a small backup power system should an electrical failure occur. The failure mode can typically be reversed by manual operation locally in the field, if required. The failure position on pneumatic or hydraulic actuators can also be selected to fail open and closed through one full valve cycle, requiring additional spring or accumulator energy depending on the number of fail positions, differential pressure, and the required direction of flow. As the number of required fail-safe mode valve cycles increase, in either full or half increments, the size of the spring and its cylinder or the accumulator will typically increase proportionately to be able to provide the additional energy required for longer operational periods. All spring or accumulator sizing should be provided by the actuator manufacturer based on the end user's operational requirements.

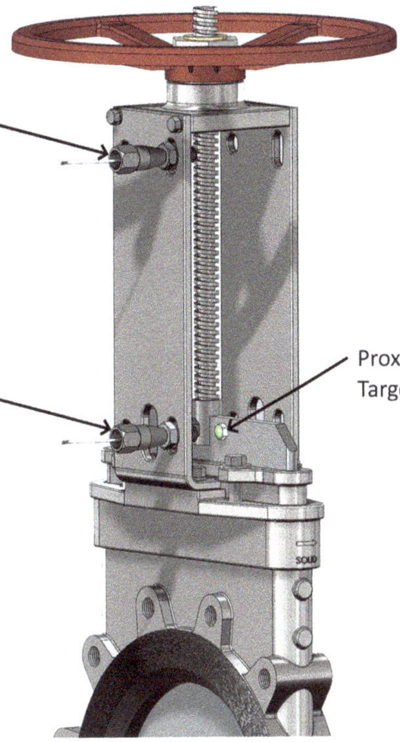


*Source*: Elite Valve

**Figure 4-5 Proximity sensor for manual or automated knife gate valves**

## POSITION INDICATION

Limit switches or proximity sensors can be mounted on manual or automated knife gate valves to indicate linear open or closed position of the gate (see Figure 4-5). Most knife gate valve manufacturers can accommodate linear feedback positioning for the full travel of the gate during manual operation. Position indication can also be provided for end of travel and/or mid position using mechanical switches. Should 0–100 percent monitoring be required, feedback potentiometers or 4-20 mA transducers can be provided without the use of an electro-pneumatic positioner. For fully automated knife gate valves, position indication and feedback are generally provided through the valve's actuator controls, but this can also be accomplished with proximity sensors.

## CHAINWHEELS

Chainwheels are used instead of handwheels or bevel gear actuators when the installed valve is not reachable from the floor or catwalks; also, sometimes they are used for large manual knife gate valve operation, where insufficient space would limit or prevent the valve's operation given other types of dimensionally larger actuation. For safety, the user should ensure that the manufacturer's chainwheel has a retention mechanism to prevent the wheel from falling from heights if the torque dislodges pins or keys.

## FLUSH PORTS

Flush ports may be used where there is a requirement to dewater the inside of a closed valve. Water under pressure can also be jetted into the valve's interior through the flush ports to clear the valve's seating surfaces, typically on an as-needed basis. When required, flush ports can be provided by the manufacturer.

Chapter 5

# Unloading, Inspection, and Storage

## UNLOADING

All valves should be carefully unloaded by gently lowering them from the truck to the ground while taking care not to drop them. The use of pallets, crating, shrink-wrap, side-load trucks, and forklifts have simplified receipt and unloading of large quantities of valves. However, the handling of single valves or uncrated valves is still prevalent, and the proper use of rigging (slings, chains, or strapping) is critical. In the case of larger valves, forklifts under the skids or slings around the body of the valve should be used for unloading. Only hoists and slings with an adequate load capacity rating to handle the weight of the valve should be used. Do not hook hoists into or fasten chains around stems, yokes, gearing, motors, cylinders, accessories, or handwheels.

## INSPECTION AFTER UNLOADING

It is important to inspect the valves and verify that no observable damage has occurred during shipment after they have been delivered and unloaded. If any damage to the crating, packaging, or valve is observed, it should be recorded before the shipper leaves the site so that any failures during testing can be supported by the condition of the valve upon its receipt.

## OPERATION AFTER DELIVERY

Large valves should be cycled through one complete opening and closing cycle in the position in which they are intended to be installed. This process will confirm:

- ease and smoothness of operation;
- complete travel to the shutoff mechanism in both directions;
- correct opening and closing rotational direction of the stem;
- manufacturer-stated number of turns to open or close the valve; and
- gate is not binding against the body, packing gland, or bonnet.

Smaller valves may also be cycled and have their operation confirmed as previously noted for larger valves, or they can be tested after installation.

The tapped holes in the valve body should be checked against the valve's shop drawings to ensure the correct bolt size and threads per inch before ordering any hardware.

It is recommended that all valves are installed with the gate in the open position in the unlikely event that the piping, when installed, causes excessive forces to the valve body preventing the operation of the gate. Once installed, the gate should be checked for smooth, nonbinding operation.

## SPECIFICATIONS

The initial inspection should also verify compliance with the specifications, including the following:

- size and pressure class (cast in the valve body);
- direction of stem rotation to open and close the valve;
- size and shape of the operating nut;
- number of turns to open or close the valve; and
- type of end connection and body style.

Verification of end configuration requires inspection by personnel familiar with the specifications, requirements, and various end configurations available, such as flanges, wafer, lugged, and semi-lugged.

## SHORT- AND LONG-TERM STORAGE

Suggested storage procedures for location, position, and coverings are noted in the following paragraphs to ensure the performance and integrity of the valve during short- and long-term storage. Short-term storage is defined as lasting from receipt of the valve to six months before installation. Long-term storage is defined as lasting from more than six months to five years before being placed in service.

Note: Grades and curing procedures may vary the long-term characteristics of the elastomer and sealing performance of the valve. Take care to follow procedures to protect the elastomer from extreme temperatures and light. Manufacturers' storage procedures will vary and must be followed in accordance with their specific requirements to ensure the valve's warranty is maintained. The following are general requirements and not specific to any single manufacturer.

## STORAGE LOCATION

The preferred storage location is in a clean, dry, indoor, protected, and secure facility. Valves stored outside may be subject to extreme temperatures; they should be covered at all times to protect them from the elements and inspected regularly. In cold weather, valves should be checked to confirm that no water has pooled and possibly frozen in areas of the valve cavities where damage to the valve could occur. In both cold and hot weather, resilient-seated valves should be checked for cracking of the elastomeric seats, and packing should be checked to ensure that it remains soft and pliable.

## STORAGE POSITION

Valves may be stored with the bodies in the vertical or horizontal position. Take care to eliminate any excess loading on the valve stanchions or posts during storage.

### Metal-Seated Valves

Valves should be placed in the fully closed position with a thin film of anticorrosion lubricant around the seating surfaces. For valves intended for use in potable water, the anticorrosion lubricant should be certified to NSF/ANSI/CAN Standard 61.

### Resilient-Seated Valves

Resilient-seated valves should be stored in the vertical position with the gate slightly open or off of seating wedges, so the gate does not energize the seal at any point on the surface of the gate. Bare and exposed metal seating surfaces should be coated with a thin film of anticorrosion lubricant. For valves intended for use in potable water, the anticorrosion lubricant should be certified to NSF/ANSI/CAN Standard 61.

### Pneumatic-Actuated Valves

For optimum protection of the cylinder, it should be stored with the valve fully opened in the vertical position. Coverings for the valve to protect the valve orifice are essential. Protective coverings for the cylinder shaft are also recommended.

For alternate storage for valves with 6-in. diameter cylinders or less, it is recommended that both the cylinder and valve body be stored in the horizontal position.

Ensure that any cylinder porting is properly plugged and sealed to prevent environmental contamination from debris or insect nesting into the cylinder chambers.

### Electric-Motor Actuated

Electric actuators should be stored indoors in a clean, dry environment where the atmospheric conditions are constant and controlled within the manufacturer's required tolerances.

Manufacturers ship actuators with temporary plugs, which may be subject to moisture migration past the plug and into the actuator body. It is essential that the manufacturer's recommended storage procedures for the valve and actuator be followed.

Most actuators have heaters and thermostats, which are not functional when the actuator is without a source of power (i.e., not energized). Once the actuator is installed or when it is stored for longer periods (and depending on the environment), power to the actuator should be maintained to prevent condensation from occurring within the actuator's enclosure.

## FLANGE COVERS

Valve flange covers should be provided and installed to protect the flange faces along with bolt hole caps to protect all threaded holes in the valve body. Bare and exposed metal flange faces should also be coated with a thin film of anticorrosion lubricant for additional protection. For valves intended for use in potable water, the anticorrosion lubricant should be certified to NSF/ANSI/CAN Standard 61.

## STORAGE INSPECTION

For indoor storage, inspection is suggested on a monthly basis with the results being recorded and traceable to each piece of equipment. Inspection on outdoor valves should be conducted weekly and the results recorded. The following is a suggested checklist when performing visual condition inspections on stored equipment:

- packaging (condition, damage, discoloration, etc.);
- coverings (secured, unsecured, damaged, etc.);
- flange protectors (adequately attached and centered, warped, damaged, etc.);
- dryness, moisture, and cleanliness; and
- corrosion.

Chapter 6

# Installation and Testing

To ensure that the valve will operate properly, reasonable care is needed in handling and installation. This chapter provides fundamental information for the installation of knife gate valves (see Figure 6-1), which will include discussions on training, valve orientation, direction of flow, bolt torqueing sequence, packing adjustments, large valve installations, and examples of failures from incorrect installations. The installation instructions from the manufacturer provided with the valve should be carefully reviewed and followed before off-loading and handling of the valves. The operation of the valve before pre-installation inspection is generally not recommended.

*Source*: DeZURIK Inc.

Figure 6-1 Knife gate valves in use at wastewater plant

# INSTALLATION TRAINING

Before installation, comprehensive valve training should be completed. This is typically provided by the manufacturer to all parties responsible for the handling and installation of the valves. Training should only be performed by a factory-authorized representative.

In addition to the installation training, it is suggested that operational training be provided by the valve and/or actuator manufacturer before connecting the valve actuator and commissioning of the valve. Operation of the valve should not be done unless the installation has been signed off by the valve and/or actuator manufacturers' authorized representatives.

## Pre-Installation Inspection

At the job site before installation, the valve should be visually inspected for any foreign material in the valve cavity. If any material is discovered, it should be removed and the body cavity wiped clean. If possible, operate the valve to expose and inspect the seating surfaces for damage before installation. Inspect the mating flange surfaces on the valve and corresponding pipe and remove all debris that could cause leakage or damage to the flange faces and gasket.

The nameplate information on the valve should be verified to ensure that the valve coincides with what the system designer has specified. It is also suggested that the valve sealing requirements, flow direction under normal operating conditions, and the pipe pressure test requirements be verified at this point in the inspection.

If the "root mean square" finish of the valve flange face or mating pipe flange face is damaged, contact the valve manufacturer for further instructions.

## Gasket Inspection

Ensure that the flange gasket surface is clean, free of debris, and in generally good condition. Inspect all gasket surfaces for damage. Verify the gasket material is as specified by the system designer. Reconfirm the type of gasket and that it is dimensionally correct to mate with the supplied pipe and valve flanges.

## Pre-Water-Up Inspection

If the valve is unidirectional seated, ensure that the valve body is installed with the flow arrow pointing in the pipeline's normal direction of flow. Confirm that all environmental or protective operational shields/guards are in place and properly installed on all automated valves.

Confirm all flange hardware is in place, properly installed, and torqued to the values based on the gasket material and valve manufacturers' recommendations. Flange hardware torques should be obtained from the gasket manufacturer based on the following information:

- flange material,
- flange class and pressure rating,
- flat- or raised-face surfaces,
- flange surface finish,
- bolt material,
- bolt diameter and thread type,
- gasket material, and
- gasket pressure rating.

Valve and actuator operation and/or commissioning procedures should be reviewed by the manufacturer or manufacturer's representative and the site inspector before the operation of the valve or filling of the system. When charging the system to test pressure, slow water-up procedures should be followed with the valve in the fully open position. Overtightening of the flange bolts in the chest area of bonnetless knife gate valves can cause seizing/binding of the gate and/or packing leakage.

## INSTALLATION ORIENTATIONS

The valve's orientation when installed in a horizontal pipeline will have a significant effect on the valve's performance and longevity (see Figure 6-2). When the stem is on plane with the springline of the pipe or at an angle below the gate, process media can build up in the body cavity. As the gate cycles, it can pull the process media into the packing or shaft seals; this can cause the valve to leak outside of the pressure boundary, requiring additional resources for unscheduled maintenance. This can also cause premature wearing of the valve's seat and gate with repeated valve cycling. Some manufacturers offer shaft-seal scrapers to mitigate this issue. If a valve has to be installed in the above-described manner, the end user should contact the manufacturer for further guidance.

Knife gate valves may be installed in vertical or horizontal pipelines (see Figure 6-3 and Figure 6-4) or at the bottom of dry or slurry hoppers. Consideration needs to be given as to whether the seat should be facing up or facing down. The location of the seat in vertical installations is largely dependent on the direction of flow and the requirements surrounding valve leakage. Unidirectional valves may need to be installed with the seat upstream to keep media from accumulating on the seat. Valves larger than 20 in. (500 mm) installed in this orientation may need external gate supports to permit smooth operation of the gate during opening/closing strokes. The end user should consult with the manufacturer or the manufacturer's representative for guidance on vertical pipe installations.

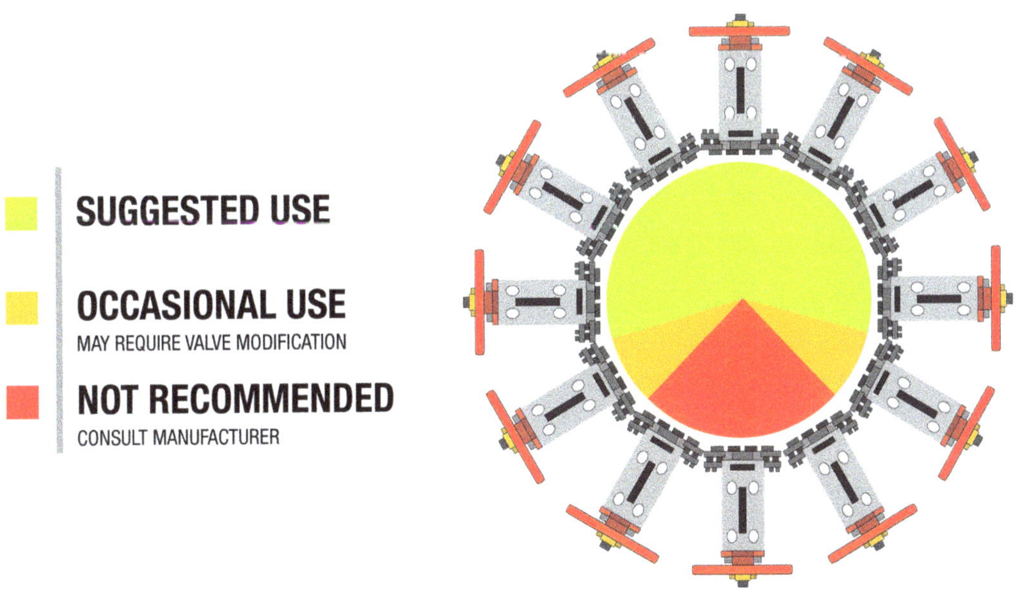


*Source*: Elite Valve

Figure 6-2 Knife gate valve suggested orientations in horizontal pipelines

*Source*: Elite Valve

**Figure 6-3 Installed slurry knife gate valve with cylinder actuation**

*Source*: Elite Valve

**Figure 6-4 Installed tapping outlet knife gate valve**

## PREFERRED DIRECTION OF FLOW

Knife gate valve designs often have a preferred direction of flow toward the gate. This type of knife gate valve is typically referred to as a unidirectional valve since it is designed to provide a seal in only one flow direction. During reverse or backflow conditions, the valve will not provide an adequate seal and leakage will occur. The amount of leakage depends on system pressures and the inherent design of the valves' affected components. Unidirectional valves will form a better seat seal with the assistance of system or process pressure. However, when a unidirectional knife gate valve experiences a reverse, excessive high pressure or backflow condition or is installed in the wrong flow direction, the gate of a unidirectional valve will push away from the seat. Some manufacturers design wedge blocks or retaining tabs into the lower portion of the body to limit the amount of gate deflection during these situations. Permanent damage to gate and valve can occur if these are not supplied.

Unidirectional valves typically have an arrow or a "seat side" designation indicating the preferred direction of flow on the outside of the body. This manufacturer recommendation should be followed and confirmed during installation of the valve. Failure to do so can compromise the sealing capability of the valve or, in extreme cases, can cause the valve gate to bend where permanent damage to the valves' affected components could occur. In Figure 6-5, the valve on the left has jams/wedges that prevent the gate from deflecting during a reverse or backflow condition; the valve on the right shows the movement of the gate in designs that do not incorporate wedges or tabs to assist in retaining the lower radius of the gate against the valve's seat in a reverse or backflow condition.

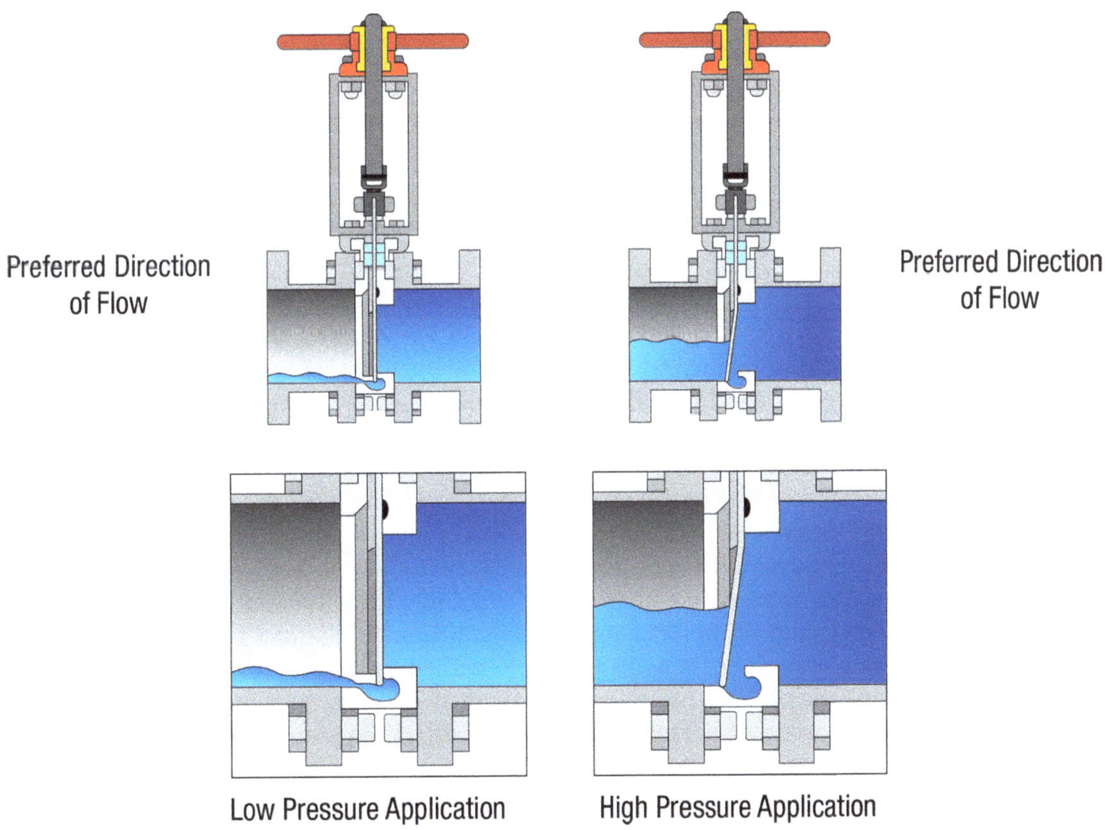


*Source*: Elite Valve

**Figure 6-5 Unidirectional knife gate valve in the reverse flow condition**

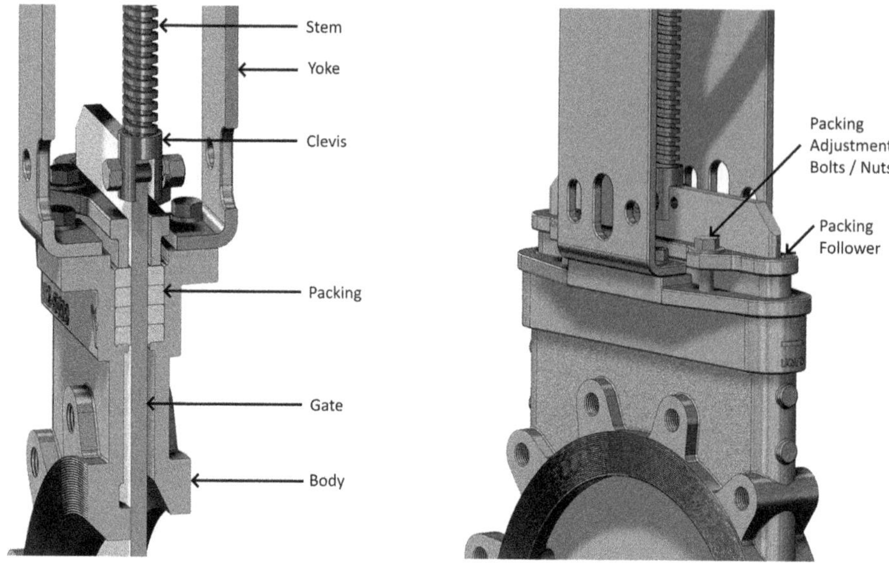


*Source*: Elite Valve

**Figure 6-6 Traditional packing gland configuration**

## BOLT TIGHTENING

Flange bolts should be tightened and torqued according to the fastener hardware and gasket supplier's recommendations—following a crisscross pattern for circular flanges and a crosswise pattern for square flanges.

Care should be taken to not overtorque the bolts in the packing/chest area. Overtightening may distort the valve body and cause leakage, gate galling, or permanent damage to the body.

## PACKING ADJUSTMENTS

When bonnetless knife gate valves are initially pressurized, the packing commonly needs to be adjusted (see Figure 6-6). Over time, the transverse gate packing may begin to leak; should leakage occur, the packing would require adjustment. The amount of adjustment necessary to stop the leakage depends on many factors, including the type of seal material used, the roughness of the gate, and how frequently the valve is cycled.

Sealing adjustments in a traditional packing gland are typically performed by tightening down the packing gland follower until leakage stops. The packing gland follower may have numerous adjustment screws or studs; they should be evenly tightened in a crisscross pattern, one-quarter turn at a time. If the bottom ring of packing extrudes below the bottom of the packing chamber during adjustment procedures, tightening of the packing may not stop leakage and continued tightening may cause the gate to bind. If replacing the packing rings does not rectify the leakage problem and packing extrusion continues to occur, an antiextrusion plate or ring is suggested.

Valves with injectable packing are adjusted by inserting pliable packing material through the access bolts on either side of the valve (see Figure 6-7). Tightening of the plugs forces the pliable material deep into the packing cavity, increasing the sealing pressure and consequently preventing or stopping any leakage.

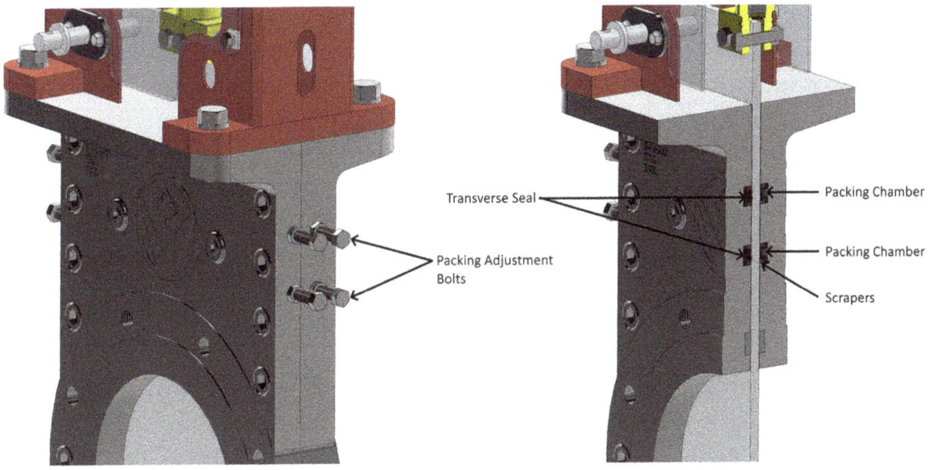


*Source*: Elite Valve

**Figure 6-7 Transverse seal packing configuration**

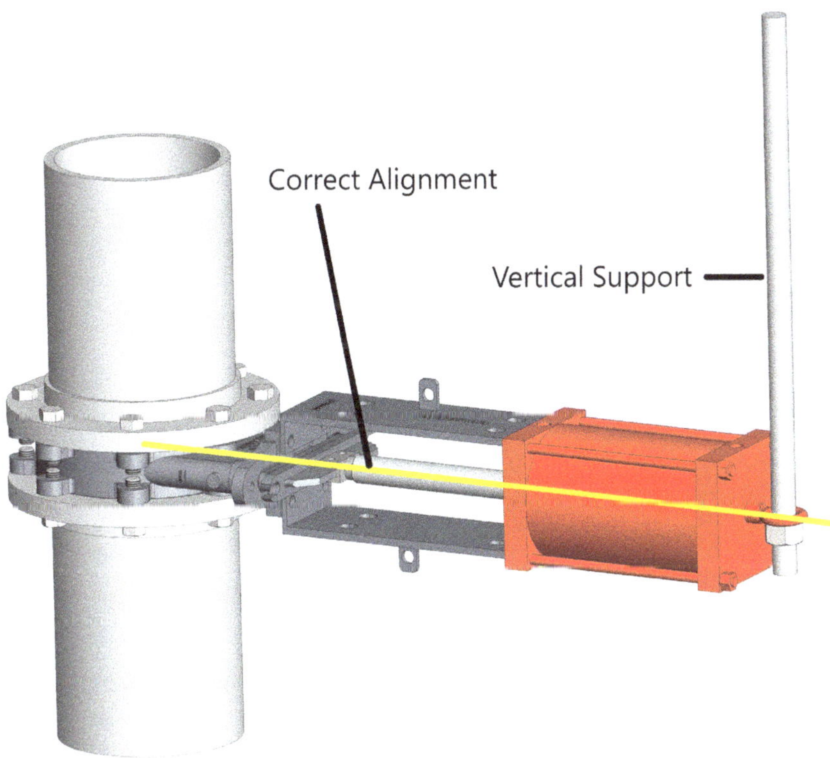


*Source*: Elite Valve

**Figure 6-8 Example of properly supported installation on each side of the valve**

## LARGE VALVES

The installation of large-diameter knife gate valves requires special considerations given that the weight of the valve may impose additional stresses on the adjacent piping system (see Figure 6-8). The installation of the valve should be designed to not act as a support for the adjacent piping, as the weight of the fluid-filled piping can cause undue stresses on the connection, mating piping, and valve—adversely affecting the operation of the valve.

Before installation, the pipework needs to be properly supported on each side of the valve. In flanged pipe connections, the weight of the valve should be able to hang from the pipe flanged connection without causing any additional stresses on the adjacent piping.

For large-diameter valves installed in a vertical pipeline where the knife gate valve will rest in a horizontal position, additional supports should be considered to support the valve. The weight of a cylinder or other actuating device acts as a cantilevered weight that transfers undue stresses to the valve body. It can also cause the gate to deflect and bow, resulting in the inability of the valve to create a proper seal or impede the operation of the gate because of binding with the valve body. It is recommended to vertically support the end of the cylinder or electric motor to mitigate these issues. The manufacturer should be consulted if support is needed. The support design will likely depend on adjacent structural features such as walls, beams, flooring, or catwalks.

## VALVE MISAPPLICATION AND MAINTENANCE

There are various types of applications that knife gate valves can be used in because of the various seat types and seals available. The process fluid, including temperature and surrounding environment, should be understood in order to supply the proper valve for a given application. The more common reasons for knife gate valve failures are installation of unidirectional valves in the reverse direction, misapplication, and lack of maintenance.

### Misapplication

A knife gate valve needs to be used in an application that it has been designed for. Typical failures from improper application can include repurposing a valve for a different application than the designer intended. Another failure can occur when a knife gate valve is used for throttling service, where the valve has not been designed to accommodate this type of application.

### Lack of Maintenance

Knife gate valves require regular maintenance and inspection to ensure proper operation and longevity. More commonly, the packing around the gate is susceptible to leakage, especially in applications where there are high differential pressure ranges. Packing failure can occur if regular inspections are not performed.

## WITNESSED TESTING

Before being put into service, all valves should be witness tested before installation to ensure proper functionality and acceptable leakage rates as identified in AWWA C520. Testing results should be recorded and traceable to each piece of equipment that underwent testing. Testing can be performed at the factory or a capable facility near to where the valve is to be installed. Typically, end users or their representatives will witness testing. End-user representatives are typically either third-party testing agencies or consultants deemed acceptable by the end user.

## PIPING TESTS

In some cases, depending on system design requirements, piping can be pressure tested to conditions beyond the design capabilities of the valve. This may cause an unacceptable amount of valve leakage to occur, damage to the valve components, or valve failure should the test conditions exceed the design pressure capabilities of the valve.

## FIELD COMMISSIONING – DRY

After installation, but before filling the system with water, the entire operating system of the facility can be precommissioned. This is typically performed in the completely dry condition; it is done to finalize control center configurations, confirm communication protocols, and ensure all equipment is in proper working order before taking the system live. This can also be an opportunity to confirm that the equipment has been installed according to the engineering designs.

A dry commissioning test does not necessarily provide a positive assurance that the installed valve will function according to expectations based on the manufacturer's design criteria. For this, a wet commissioning test must be performed.

## FIELD COMMISSIONING – WET

After installation of the equipment and field commissioning in the dry condition, the entire operating system of the facility can be commissioned. When practical, this is performed under full system pressure to confirm equipment performance under operational conditions based on the manufacturer's stated design criteria.

This page intentionally blank.

Chapter 7

# Operation, Maintenance, and Safety

## GENERAL

To ensure safe and reliable operation of the knife gate valve/actuator assembly and the piping system, reasonable care is needed in the handling, installation, and maintenance of the equipment. This section provides safety concerns and maintenance procedures for manual and power-actuated knife gate valves. When using power actuators, the instructions provided with the actuator should be carefully reviewed and followed so as to not permanently damage the equipment.

## SAFETY CONCERNS

Before installation, inspection, or servicing, the local, state, provincial, or federal safety procedures should be reviewed and established. Ensure anyone working on the knife gate valve has been trained on the specific valve and is aware of the manufacturer's most recent or applicable installation and maintenance manual for the valve being worked on or installed. The instruction manual supplied by the manufacturer should be reviewed in detail before commencing maintenance on any style of knife gate valve.

Review and follow all local, state, provincial, or federal confined space regulations when accessing the valve for inspections or maintenance. All precautions must be considered for a safe work environment and operation of the valve or actuator assembly, and all risks must be evaluated and mitigated based on the planned work to be conducted.

Do not disassemble any components of the valve without first verifying that the flow and pressure is at zero within the adjacent piping while performing any work inside of valve chambers.

## MAINTENANCE PROCEDURES

The manufacturer's manual will contain a list of routine preventive and corrective maintenance procedures specific to the equipment purchased. Preventive maintenance procedures will reduce the likelihood of the knife gate valve seizing, jamming, or leaking. These should be followed to prolong the service life of the valve. Maintenance procedures typically will include performing regular visual inspections, a scheduled cycling program to confirm valve operation and performance, and regular lubrication of parts in rubbing contact with each other.

## PREVENTIVE MAINTENANCE SCHEDULES

Maintenance schedules will vary based on the type of service and frequency of use. Automated valves which cycle regularly will have shorter intervals between packing gland adjustments and other required maintenance items. Consult the manufacturer's manual for the particular requirements of the supplied valve and actuator.

## RECORDKEEPING

Keep a detailed record of valve maintenance to ensure safe and reliable system operation and continuity of the preventive maintenance program. Digital recordkeeping at the completion of any maintenance is the fastest and most secure method to ensure records are current, safe, and readily viewable by others.

# Index

Note: *f.* indicates figure; *t.* indicates table

## K



## L



## M



## O



## P

## R



## S



## T



## U



## V



## W



## Y

This page intentionally blank.

# AWWA Manuals

M1 *Principles of Water Rates, Fees, and Charges,* #30001
M2 *Instrumentation and Control,* #30002
M3 *Safety Management for Water Utilities,* #30003
M4 *Water Fluoridation Principles and Practices,* #30004
M5 *Water Utility Management,* #30005
M6 *Water Meters—Selection, Installation, Testing, and Maintenance,* #30006
M7 *Problem Organisms in Water: Identification and Treatment,* #30007
M9 *Concrete Pressure Pipe,* #30009
M11 *Steel Pipe—A Guide for Design and Installation,* #30011
M12 *Simplified Procedures for Water Examination,* #30012
M14 *Backflow Prevention and Cross-Connection Control: Recommended Practices,* #30014
M17 *Fire Hydrants: Installation, Field Testing, and Maintenance,* #30017
M19 *Emergency Planning for Water and Wastewater Utilities,* #30019
M20 *Water Chlorination/Chloramination Practices and Principles,* #30020
M21 *Groundwater,* #30021
M22 *Sizing Water Service Lines and Meters,* #30022
M23 *PVC Pipe—Design and Installation,* #30023
M24 *Planning for the Distribution of Reclaimed Water,* #30024
M25 *Flexible-Membrane Covers and Linings for Potable-Water Reservoirs,* #30025
M27 *External Corrosion Control for Infrastructure Sustainability,* #30027
M28 *Rehabilitation of Water Mains,* #30028
M29 *Water Utility Capital Financing,* #30029
M30 *Precoat Filtration,* #30030
M31 *Distribution System Requirements for Fire Protection,* #30031
M32 *Computer Modeling of Water Distribution Systems,* #30032
M33 *Flowmeters in Water Supply,* #30033
M36 *Water Audits and Loss Control Programs,* #30036
M37 *Operational Control of Coagulation and Filtration Processes,* #30037
M38 *Electrodialysis and Electrodialysis Reversal,* #30038
M41 *Ductile-Iron Pipe and Fittings,* #30041
M42 *Steel Water-Storage Tanks,* #30042
M44 *Distribution Valves: Selection, Installation, Field Testing, and Maintenance,* #30044
M45 *Fiberglass Pipe Design,* #30045
M46 *Reverse Osmosis and Nanofiltration,* #30046
M47 *Capital Project Delivery,* #30047
M48 *Waterborne Pathogens,* #30048
M49 *Quarter-Turn Valves: Head Loss, Torque, and Cavitation Analysis,* #30049
M50 *Water Resources Planning,* #30050
M51 *Air Valves: Air-Release, Air/Vacuum, and Combination,* #30051
M52 *Water Conservation Programs: A Planning Manual,* #30052
M53 *Microfiltration and Ultrafiltration Membranes for Drinking Water,* #30053
M54 *Developing Rates for Small Systems,* #30054
M55 *PE Pipe—Design and Installation,* #30055
M56 *Nitrification Prevention and Control in Drinking Water,* #30056
M57 *Algae: Source to Treatment,* #30057
M58 *Internal Corrosion Control in Water Distribution Systems,* #30058
M60 *Drought Preparedness and Response,* #30060
M61 *Desalination of Seawater,* #30061
M62 *Membrane Applications for Water Reuse,* #30062
M63 *Aquifer Storage and Recovery,* #30063

M65 *On-Site Generation of Hypochlorite,* #30065
M66 *Cylinder and Vane Actuators and Controls—Design and Installation,* #30066
M68 *Water Quality in Distribution Systems,* #30068
M69 *Inland Desalination and Concentrate Management,* #30069
M75 *Elastomers for Waterworks: Pipes, Valves, and Fittings,* #30075
M77 *Condition Assessment of Water Mains,* #30077

www.ingramcontent.com/pod-product-compliance
Ingram Content Group UK Ltd.
Pitfield, Milton Keynes, MK11 3LW, UK
UKHW052014041026
15590UKWH00008B/2322

9 781647 17027